Endorsements

"Darw

"The need for a super-intelligent force to create and sustain living things is well set out and without question. I also appreciated all the research that was done to demonstrate the historical importance and recognition of God in the nation lives of the four nations you selected." -- George Matzko, Ph.D.

"*Darwin's Replacement* is a well-written review of the enormous complexity of all life written from the perspective of its molecular biochemical foundation. The complexity is not only amazing but elegant. Most people enjoy learning about amazing feats, thus the popularity of 'Ripley's Believe It or Not' and similar books. Our body and its complexity are so familiar to us and it works so well for most of us that we often take it for granted.

Mr. Rogers' book helps us to realize that we are all walking miracles; and understanding how it functions is both awe-inspiring and helps us to appreciate the body we all inhabit while on this Earth." -- Jerry Bergman, Ph.D.

"*Darwin's Replacement* provides credible assessment of the weaknesses within mainstream evolution theory, and proposes a reasonable, evidence-based alternative for the origin and development of life. -- Nicholas Comninellis, M.D.

"ATOMIC BIOLOGY promises to restore the true foundations of science back to the realm of observation. Current forays into metaphysical speculation and presupposition by 'experts' seem to have caused division, confusion, and misunderstanding in the scientific conversation." -- Jack Taylor, Ph.D.

“My recommendation would be to rework the book as a series of 1-to-2-page studies for adult Sunday school classes and/or Christian High School classes. It could be useful as an introduction to the topics, in that format.” -- David Snoke, Ph.D.

Your book demonstrates the amazing complexity of life, starting with even the simplest cell, and the numerous conditions needed to sustain life. That all this could be the result of blind, random evolution is highly implausible, and statistically, virtually impossible. Hence, as your book concludes, this points to a super-intelligent Creator. Your book also notes that the USA, the UK, Canada, and Australia were all founded on submission to the Christian God, and urges those countries to return to acknowledging God, also in science classrooms. I heartily agree with all this.”

-- John Byl, Ph.D.

“There are only two possibilities for the existence of life: accidental or purposeful. Using science and mathematics, Atomic Biology proves beyond a shadow of doubt that life cannot be accidental. Then the book shows that the only being capable of the creation of life and its orchestrated maintenance is the historical Omniscient, Omnipotent, and Omnipresent Triune God of the Bible and our Nation.” -- Sharon E. Cargo, D.V.M.

“Hi, (Reality R&D). I recently bought the book and am reading it slowly and out loud to myself so that it sticks, but can I say that when I heard Mr. Rogers on Vision I knew that this was a book I’d been waiting for a very long time. I honestly can’t put into words how exciting this is for me to finally have something I can refer to when discussing creation and not sound like a loony.”

-- Linda Houston, Australia

DARWIN'S REPLACEMENT - SUMMARY INTRODUCTION

HOW WE ARE AMAZINGLY MADE

AND CARED FOR BY OUR SUPER-INTELLIGENT CREATOR, THE GOD OF OUR NATION

Dedication

To Our Super-Intelligent and Caring Creator,

Credit for Life Where Credit Is Due.

Acknowledgements

Our respect and gratitude go to the forty-five scholars whose work has helped in the discovery and development of this God-based life-science called "Atomic Biology." Our Creator has probably been using this science since the beginning.

We are also grateful to the encouragers, endorsers, and supporters who have helped to keep the work moving forward.

A Gift For Others

"Always be prepared to give an answer to everyone who asks you to give the reason for the hope (belief) that you have. But do this with gentleness and respect …" 1 Pet. 3:15

This booklet can help you greatly towards fulfilling this commission in your personal and professional life. God bless you!

LRG

LIFETIME REFERENCE GUIDES INC.
P.O.Box 51613 RPO Park Royal
West Vancouver, BC, Canada V7T 2X9
www.lifetimereferenceguides.com

Cover design by Arneeon.com with Shutterstock images.

ISBNs
978-1-9992097-8-0 (eBook)
978-1-9992097-7-3 (Paperback)

Printed in Canada and USA

Biology
Science
Education
Government

Contents:

Introduction:

Have you ever thought of all the amazing and careful works our Creator performs for you every day? It is worth a pause, just to smell a rose, or watch and hear some birds, or hug a friend, or eat some fruit and enjoy good food.

For most of us, these things are so easy and reliably available, we usually take them for granted, but none of this just happens. Each "Life" takes a huge amount of caring work by our Creator, the God of our nation.

The study of this brilliant work is a new subject called "Atomic Biology," and it is fascinating.

This booklet provides a simple introductory study for teachers, students, and congregants.

Hopefully, most teachers and professors want to provide accurate information to their students. Ideally, more education leaders will insist that best explanations are taught, especially in the sciences. This means allowing teachers and students to "follow the evidence wherever it leads and without restriction or reprisal." Currently, teaching in public schools, colleges and universities, that evolution-only is the cause of life, disallows teaching the truth that super-intelligence is essential to construct and sustain our cell parts, cells, and us.

Conscientious educators are needed to fix this serious problem that has far-reaching effects on personal lives. What teacher wants to be forced to teach their students a big lie.

"If it could be demonstrated that any complex organ could not possibly have been formed by numerous successive slight modifications, my theory would absolutely break down." [1]

-- Charles Darwin

This summary will show:

1. The absolute essentiality for Super-Intelligence, far beyond humankind's level, for building and sustaining all living cell parts, cells, and entities, including us;
2. The enormous amount and speed of the super-intelligent works performed for us in our bodies every second of every day;
3. The phenomenally careful precision of the brilliant construction works using atoms in building each of our cell parts, cells, and the total "us."

These are all intelligent works that "evolution" is incapable of performing because, by definition, it has no intelligence to use.

For many reasons that you will see in this introductory booklet, evolution should be ruled out as the true origin and cause of life.

A brief video is at: www.atomicbiology.com/abvideo

References:

1 Darwin, Charles, *On the Origin of Species by Means of Natural Selection,* 1st edition, 1859, p. 189, John Murray, London, Eng. available online from Darwin-online.org.uk.

Chapter 1: How We Are Amazingly Made

Charles Darwin theorized in his 1859 book, *On the Origin of Species,* that a common ancestor existed for all living things. However, to this day there has been no plausible explanation for how that necessarily complex original ancestor came into existence without intelligent works with atoms. The construction of even the simplest cell is now known to be so utterly complex that a growing number of evolutionists disagree with Darwin's theory.

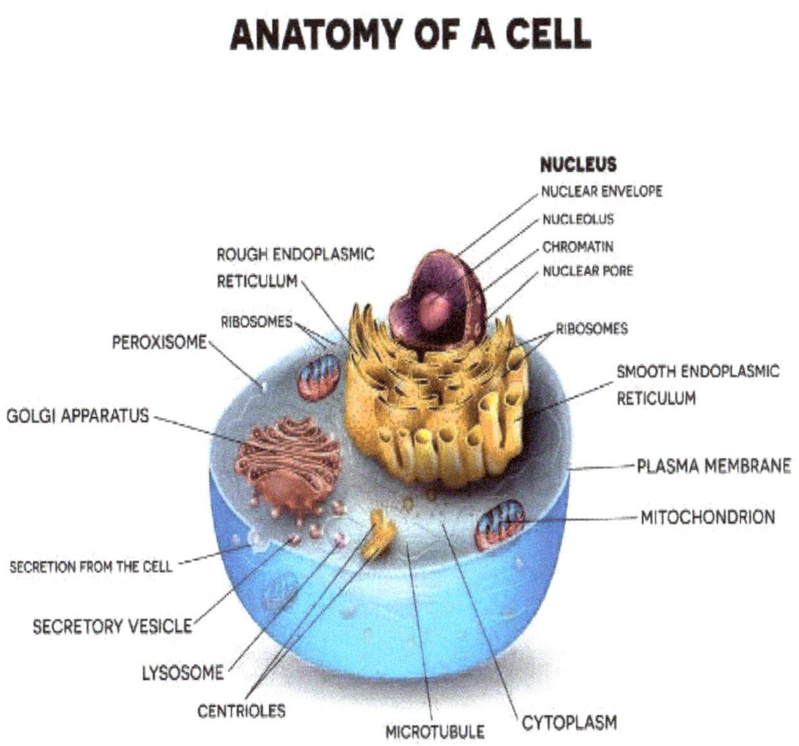

Shutterstock – Tefi

It is the brilliant design and construction of cell parts, cells, and living entities like us, that has compelled scientists to move away from their former belief that "evolution" created all life.

Here are some of the many amazing factors that illustrate the brilliant engineering, design, and construction works for life by a super-intelligent and caring entity:

* A 150-pound male is built with about 100 trillion constructed cells including over 200 cell types;
* Each one of our approximately 20 trillion red blood cells is constructed with about 280 million hemoglobin molecules,[1] and each of those molecules is built with about 10,000 correct atoms.[2] So, each one of our red blood cells consists of about 2,800 billion correctly selected, counted, and assembled atoms;
* About 2.3 million new red blood cells are produced every second,[3] 24/7, to replace those that are worn-out in this average 150-pound male (i.e. about 6400 quadrillion correct atoms/second). You can estimate the number of atoms per second that are being selected and assembled to make your red blood cells, using your relative weight. The constant replacement of our blood cells and almost every other cell type in our body is a huge and complex task performed for every human every second of every day by our caring Creator.
* Our many cell organelles must be carefully constructed to exacting specifications. For example, the four DNA bases made for all life forms are constructed precisely to the following formulae:

Adenine (A) -	chemical	formula:	$C_5H_5N_5$
Guanine (G) -	"	"	$C_5H_5N_5O_1$
Cytosine (C) -	"	"	$C_4H_5N_3O_1$
Thymine (T) -	"	"	$C_5H_6N_2O_2$

Notice how carefully the atoms must be selected, counted, and assembled. One atom off is likely lethal. For Adenine, $C_5H_5N_5$ is required. $C_5H_5N_6$ will not work, nor will $F_5H_5N_6$ or any other mix.

The required atoms are picked up from our digestive system and carried in our blood stream to where they are needed for constructing all our new cell parts. For DNA, over three billion pairs of these bases must be precisely assembled and arranged in various special sequences (codes) to construct the hardware and assemble the functioning DNA and RNA software instructions required to help operate up to 40 different molecular machines in each one of most of our cells;

* How about our amazing eyesight?

Even Mr. Darwin said, "*To suppose that the eye with all its inimitable contrivances for adjusting the focus to different distances, for admitting different amounts of light, and for the correction of spherical and chromatic aberration, could have been formed by natural selection, seems, I freely confess, absurd in the highest degree.*[4]

Then he wrongly speculated on how it might just happen without intelligent guidance.

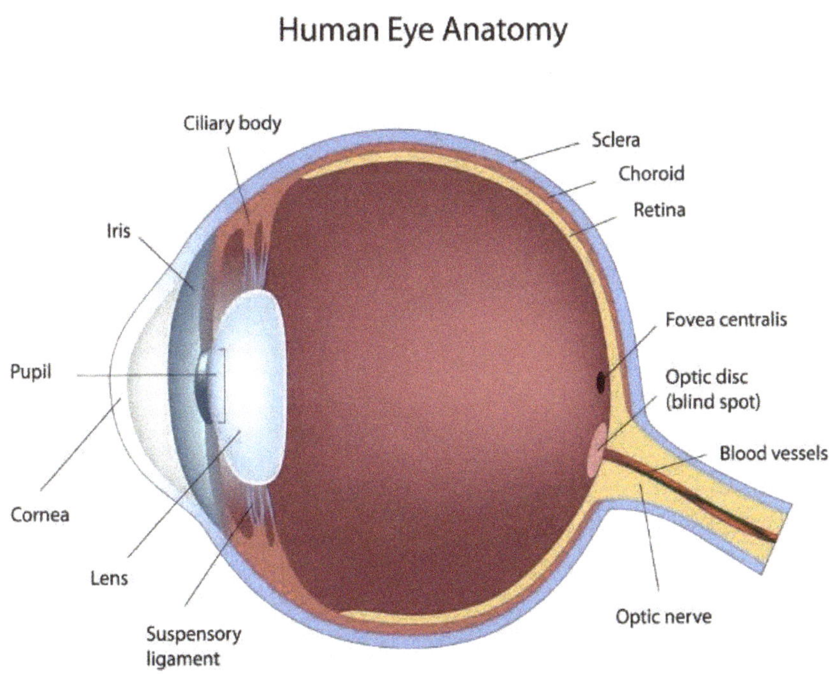

Alila Medical Media/Shutterstock

* Then there are all our other amazing senses including hearing, tasting, smelling, and touching (awesome and reliable but so easy to take for granted).

How about our incredible brains? Prof. Paul Reber estimates our brain storage capacity is equivalent to about *three million hours of recorded TV shows.*[5] Yet evolutionists claim that our brain is constructed without any intelligent work or care.

Magnified Image of a Cell Membrane Portion

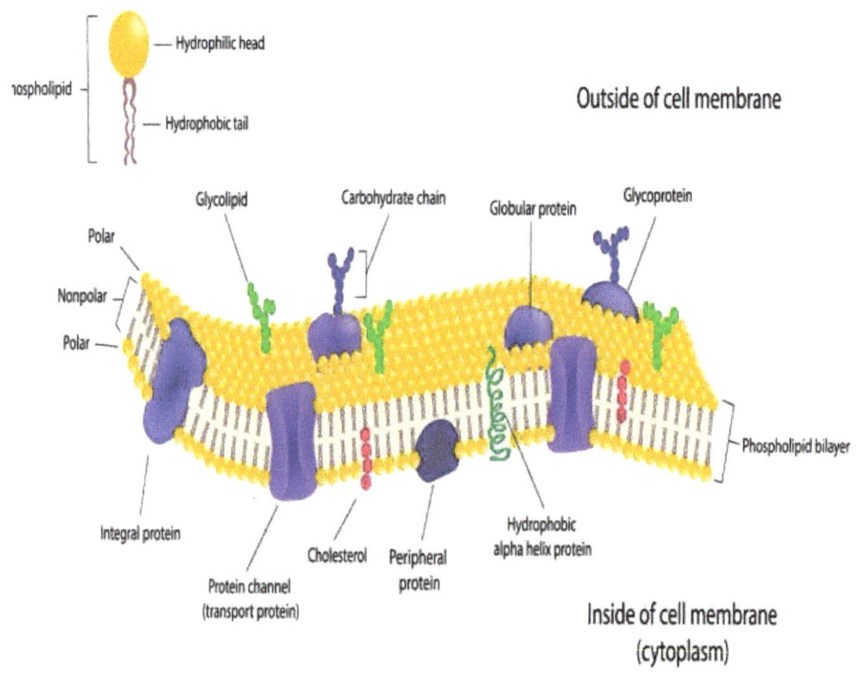

Kallayanee Naloka / Shutterstock

Notice how complex it is to build even this small portion of the membranes encasing each of our cells. A phenomenal amount of intricate works with atoms are absolutely essential.

- **Either evolution is the cause of life, or it is not.**
- **Proof now exists to show that Super-Intelligence and skills (far above the human level), are essential to construct our various cell parts, cells, and us.**
- **Evolution, by definition, has no intelligence to use, therefore, it is falsified as the cause of life.**
- **Evolution should no longer be taught as the cause of life.**
- **Unfortunately, students will still have to say "evolution does it" on your public school and university exams for a while longer, but you can also remember the above factors regarding how much your Creator cares and works for you.**
- **You usually should not challenge your teachers over this fact, but you can decide what is logically true.**

References:

[1] Tortora, G. J., *Principles of Anatomy and Physiology,* John Wiley & Sons, New York, NY, 2008.

[2] Perutz, Max, *Science is Not a Quiet Life: Unraveling the Atomic Mechanism of Hemoglobin,* World Scientific, Hackensack, NJ, 1997.

[3] Pallister, C. J., *Haematology: Biomedical Science Explained,* Butterworth-Heinemann, Burlington, MA, 1999.

[4] Darwin, Charles, *On the Origin of Species by Means of Natural Selection,* 1st edition, 1859, p. 186, John Murray, London, Eng., available online from Darwin-online.org.uk.

[5] Reber, Paul, "What Is the Memory Capacity of the Human Brain," *Scientific American,* May/June 2010, www.scientificamerican.com/article/what-is-the-memory-capacity/.

Chapter 2: Why Is It "Super-Intelligence"?

In 2016 the three Nobel Prize Winners in Chemistry, J. P. Sauvage, Sir J. F. Stoddart, and B. L. Feringa, showed, unintentionally, that mankind has nowhere near enough intelligence and skills to build even the simplest molecular machine in our cells. They received this prestigious prize after 33 years of work to make a few simple molecular machines. The best they could make were, relatively speaking, almost infinitely more simplistic than the simplest ones built for our new cells every day.

This shows that "Super-Intelligence," far greater than mankind has, is essential for building the molecular machines for our cell parts, cells, and ultimately the total *us*.

The evolutionists who say that "natural mechanisms" do this work, invite a much larger problem for their theory: these "natural mechanisms" would first have to be constructed with far more intelligence and better equipment than the Nobel Prize Winners had and work them within the microscopic confines of a cell.

The only *Super-Intelligent Cause* known to mankind is the God of our Nations (USA, UK, Australia, and Canada). The God of "In God We Trust," "One Nation Under God," "God Save The Queen," "God Keep Our Land Glorious and Free," the God of "Christmas", "Easter" and "Thanksgiving Day" -- *that* God.

He constructs our cell parts, cells, and our body using atoms from our foods which He made with the same atoms He found in the dust or soil in gardens, fields, and orchards.

"…for dust you are and to dust you will return." Genesis 3:19 NIV

"For you created my inmost being; you knit me together in my mother's womb. I praise you for I am fearfully and wonderfully made; …" Psalms 139:13-14 NIV

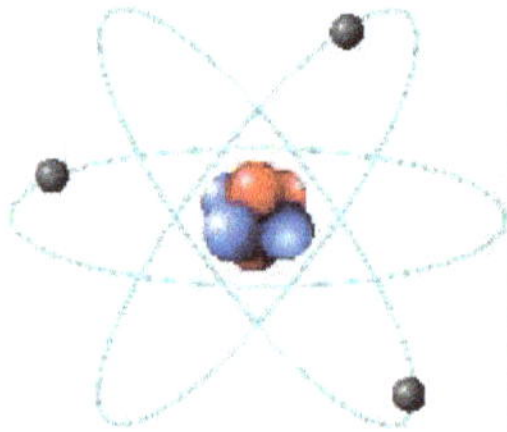

Shutterstock/Azure

Van Wedeen and L.L. Wald of the Martinos Center for Biomedical Imaging Human Connectome Project, state that "The brain's many regions are connected by some *100,000 miles* of fibers called white matter, enough to circle the Earth four times."[1] Making our brains out of "dust" requires Super-Intelligence.

Neuroscientist, Harvard professor Jeff Lichtman, is studying brain compositions. He was interviewed by Carl Zimmer who wrote in *National Geographic,* February 2014, "So far the largest volume of a mouse's brain that Lichtman and his colleagues have managed to re-create is about the size of a grain of salt. Its data alone totals a hundred terabytes, the amount of data in about 25,000 high-definition movies." They also related that a mouse's brain contains about 70,000,000 neurons and a human brain contains about 1,000 times that number.[2]

Douglas Axe, Ph.D., is an engineer-turned-molecular-biologist and director of the Biologic Institute. In his recent book, "*Undeniable: How Biology Confirms Our Intuition That Life Is Designed,*" he reminds us that, "*The human brain is different.... Being the most remarkable component of the human body, it is arguably the most outstanding physical invention ever to exist.*"[3]

This phenomenal computer, our brain, had to be super-intelligently built of atoms from the soil, air, and water with God's phenomenal two-step process: atoms in the soil are built into food, then those same atoms in our food are used to construct our brain cells.

Where else could these atoms come from, and who else has the super-intelligence to assemble them in such a phenomenal manner?

Then, of course, there is the matter of adding "Life" to these inanimate atoms. (More Super-Intelligence required).

" ... the Lord God formed the man from the dust of the ground and breathed into his nostrils, the breath of life ...". Genesis 2:7 NIV

Every living cell, including all of ours, needs God's breath of life today. When that is removed from any cell, it dies.

God's "evolution" work is shown here in deconstructing a crawling worm and reconstructing the atoms into a beautiful flying butterfly in 14 days, not millions of years.

Stephen Russell Smith Photos/Shutterstock

Evolutionists might say this is just "metamorphosis," but the name does not do the work of changing a crawling worm into a beautiful flying butterfly.

References:

[1] Wedeen, Van, and Wald, L. L., in article "Secrets of the Brain" by Carl Zimmer, *National Geographic,* February 2014, p. 34.

[2] Lichtman, Jeff, in article "Secrets of the Brain" by Carl Zimmer in *National Geographic,* February 2014, pp. 39, 43.

[3] Axe, Douglas, *Undeniable,* Harper One, New York, 2016, p. 259. 259.

Chapter 3: Why Not "Evolution"?

As "evolution" has many definitions, we must define the word so the readers can all be 'on the same page.'

The definition of "evolution" that we are using at The Atomic Biology Institute, includes the theories of evolution currently being taught in our public schools, colleges, and universities, including Darwinism, Neo-Darwinism, and macro-evolution. They include the theoretical concept of a "universal common ancestor" formed by a chance assembly of atoms billions of years ago. This imagined "common ancestor" had to be complex enough to have life, to function, find nourishment and digest that for survival, and to reproduce with improvements with all the molecular machinery necessary to enable this work to be completed.

The mathematical odds against this just happening are so enormous as to make it realistically impossible.

No one is able to adequately explain how such a complex original ancestor could have come into existence without intelligent work. In reality, there has to be a "common Designer and Builder".

For the imagined "common ancestor" to have the intelligence and ability to assemble an offspring with even more sophisticated features, is so improbable that many, if not most, rational thinking evolutionary scientists dropped their belief in a common ancestor.

The highly complex construction work that goes into building even the simplest living entities, is turning an increasing number of scientists away from "evolution" as the cause of life.

Cambridge University Ph.D., Stephen C. Meyer, stated in his book, *Signature In The Cell,* that "The simplest extant (still surviving) cell, *Mycoplasma genitalium* – a tiny bacterium that inhabits the human urinary tract – requires 'only' 482 proteins to perform its necessary functions and 562,000 bases of DNA (just under 1,200 base pairs per gene) … .

"Based on minimal-complexity experiments, some scientists speculate (but have not demonstrated) that a simple one-cell organism might have been able to survive with as few as 250 to 400 genes." [1]

Can anyone honestly believe that a living entity, even that small, could be constructed by chance using no intelligent work or direction? Let's be realistic and honest.

We have pointed to the Nobel Prize Winners' evidence showing (unintentionally) that super-intelligence is essential to build the molecular machines required for our cells to function. Evolution claims that intelligence is not required to produce life. All that is required is time, chance, and the traditionally accepted (though still debated) four forces that hold the universe together, namely, gravity, electro-magnetism, and the strong and weak nuclear forces.

In addition to no evidence of a common ancestor, and evolutionists' claim that no intelligence is necessary for building cell parts and cells, are the following factors supporting a Creator:

* The atoms our cells are made of do not have the ability to move themselves into their precise position in a cell's organelles. This requires a super-intelligent entity to find, select, count, and precisely assemble and fasten them in the proper sequence.

* The fossil record shows no evidence of clear transitional forms for the major categories of life. The few fossils claimed to be transitions are dubious at best. This was a major concern for Darwin as well.

* *"Dead dogs don't bark,"* which is to say that although all the required atoms, molecules, and cells are precisely built and placed into their correct position for eyes, ears, teeth, brain, legs, heart, and so on, *without the super-intelligent breath-of-life, those atoms, molecules and cells, or the dog, are not going to move one*

millimeter. This God-given breath-of-life is crucial to every living entity. When removed, the entity's life ends.

* All living entities are built of cells containing DNA. DNA is like a computer software program which uses a very sophisticated, highly complex coding system to assist in the multiple and varied functions that cells have to perform in order to help keep a living entity alive. Complex, intelligent, functional codes, like those in DNA, cannot be written or programmed without an intelligent coder or programmer.

* Therefore, Evolution, having no intelligence, is **falsified** as the cause of life.

* Many factors exist that must be intelligently 'tuned,' highly regulated, and crucially consistent in order for our planet to function, and for living entities to exist. *Uncontrolled, unguided conditions would quickly lead to the end of all life.*

Here are just a few conditions which life must have in order to exist:

1. Temperatures must not be too hot or too cold.
2. Sufficient water must be available.
3. Sufficient food must be constantly constructed.
4. A super-intelligent cause (God, the only possible one) is needed to build cells from atoms.
5. God is needed to "breathe life" into these cells.
6. Adequate sunlight must exist for life to exist.
7. The atmosphere must be just right.
8. Gravity must not be too weak or too strong.
9. Electricity flow must be controlled in our body.
10. A super-intelligent agent is needed to keep all of these necessary factors (and more) in coordinated balance.

Reference:

[1] Meyer, Stephen C., *Signature in the Cell,* Harper Collins, New York, NY, 2009, p. 201.

Chapter 4: God In Our Governments.

In this book, we refer to the governments of the USA, the UK, Australia and Canada.

Many decades ago, our forefathers and founding fathers intuitively understood that living entities including livestock, food, siblings, children, and themselves, were all made by an intelligent cause whom they called "God," their "Creator," and their "Provider."

They also found truth and internationally beneficial wisdom in His Word, the Holy Bible.

Out of respect and appreciation, they implemented special days of recognition for God's faithful works on their behalf including Thanksgiving Day, Christmas, and Easter.

They highly recognized God in their Declarations, Justice Systems, Constitutions, Pledges, Anthems, and more.

They put His name on public buildings, war memorials, currency, and more.

They prayed to Him in times of personal, national, and international distress then, as good leaders sincerely do today.

He is very patient and forgiving with us, but only to the point where He is purposefully disrespected or angered because praise is given somewhere else for the enormous work He performs for each of us every second of every day.

History shows what happens to people and nation who turn their focus against Him.

Our students have the right to be taught "Why God is So Highly Recognized by Their Government."

Now the life-science of Atomic Biology gives scientific reasons that back up all governments' positions of recognizing God.

Fortunately, in the USA, recognitions of our Creator and appreciation of His works are regaining prominence as states pass legislation to reinstall the national motto, "In God We Trust," back into public schools. Also, as skepticism grows against Darwinisms, some states are seeking an alternative science to teach regarding the cause of life.

We propose that "Atomic Biology" is a logical candidate to be considered as this significant Darwinism alternate.

As one who lived during the "Happy Days" of the 1950's, when God and His advice were widely taught, honesty and trust were honorable and practiced traits. Those of us who know the benefits of our Creator's influence, are aware of what our society has lost by the widening disrespect and ignoring of His works and blessings.

If we want God to bless America and other nations, we need to heed His advice and appreciate His enormous works on our behalf.

In addition to replacing the national US motto, "In God We Trust," into the halls of education, it would be good to explain "Why" to the students. Reinstalling "The Ten Commandments" at the same time would provide some lifetime wisdom and guidance for our students to live by. The 'ten' were given to us by the God of our Governments for the benefit of our people.

A major goal of The Atomic Biology Institute is to explain some of the multitude of reasons why we can trust and appreciate God and His phenomenal caring works for each one of us.

It would also be helpful for students to understand their enemy who can tempt them, pushing some who give-in to temptations, into deep trouble. A thorough understanding of how temptation can be terribly destructive, could be an important lesson for all individuals to be taught.

Often temptation can make some action look exciting, fun or instantly gratifying, but the consequences can be devastating in the short term or even for a lifetime. There are probably no more devastating temptations than the abuse of alcohol, drugs, or sex.

A brief but graphic course called "Choices and Consequences" near the beginning of every school year, could save many individuals a lifetime of grief. Governments, through their education departments, have a hugely significant opportunity to positively influence the lives, gratitude, enjoyment, personal health, ambulance costs regarding over-dosers, health-care costs, policing costs, and heartaches, for multi-millions of citizens and taxpayers.

A huge part of education is the imparting of practical knowledge and a moral compass, in addition to applied wisdom for our students. Unfortunately, evolution has greatly eroded the morals of the nations where it has been allowed to become dominant.[1]

The topics of "Choices and Consequences" will be discussed thoroughly in our textbooks.

For clarity regarding the various interpretations of "separation of church and state," we must point out that God is not 'the church'. The church can be a building or a group of people who are Satanists, New Agers, Catholics, Protestants, Mormons, Jehovah's Witnesses, cultists, or others. *No church can create any living thing*. God is a highly regarded part of some churches, just as He is a highly regarded part of our governments, but He is definitely not 'the church.'

References:

[1] See Dr. Jerry Bergman, *How Darwinism Corrodes Morality: Darwinism, Immorality, Abortion and the Sexual Revolution*, Joshua Press, Kitchener, ON, Canada; 2017; Dr. Jerry Bergman, *The Darwin Effect. Its Influence on Nazism, Eugenics, Racism, Communism, Capitalism & Sexism*. Master Books, Green Forest, AR. 2014

Chapter 5: Why Is Evolution Still Being Taught?

Evolution is still being taught because it is the enforced exclusive subject of the cause of life. No criticism is usually allowed. This is truly **anti-science** because it forcibly eliminates the basic principle of scientific discovery, i.e. encouraging exploration and following the evidence wherever it leads without restriction or reprisal.

Restrictions of this type stifle the development of the best explanations, best solutions, and best education.

These 'evolution-only' rulings should be illegal, but in reality, the opposite is currently true. Teachers and professors who have dared to suggest that there may be some intelligence in the designing of living entities, have been severely reprimanded.[1].... **more anti-science.**

However, in growing numbers, brave scientists are daring to publicly declare their skepticism of the dogma of evolution. It is bizarre that in our democracies where freedom of thought and speech have been fought for, and died for, that there should be these penalized restrictions in science, a field of such great potential for tremendous good.

The second reason is that the religious belief of Darwinism has been allowed to dictate restrictions onto other religious belief systems. This is a significant conspiracy of major negative influence on our society in general. They have been allowed to use our public educational institutions as their "churches" and teachers as their "preachers" to preach exclusively their doctrine that 'evolution-only is the cause of life – no God allowed.'

Three generations of enforced teaching of Darwinism as the cause of life, has muddied the image of the God of our nations, our Creator and food Provider on whose principles our great nations are founded.

History shows that ignoring God and His advice leads to trouble for the nations, including moral decay and the strife that comes with it, e.g. disrespect for law, depression, anxiety, addictions, family breakdown, greed, and crime **plus all the related costs.** All these problems come from rejecting God's wise advice, including particularly the Ten Commandments in His 'operator's manual,' the Holy Bible.

Instruction in "Bible literacy" would be of great benefit and value to **all** in our society.

It stands to reason why there is less strife for people who use this great advice.

Now that Super-Intelligence, (far greater than that of mankind), has been 'proven' essential for the design and construction of cell parts, cells, and living entities, including the total *us*, it is time to bring the God of our nations back to our students.

For those who say that no science can be "proven," this is primarily an evolutionist's statement as their theory cannot be proven. In fact and in truth, it is now proven to be false. Even mutant cells have to be intelligently constructed.

If you suggest to an evolutionist to jump from a 20-storey building to show that gravity is not 'proven,' he or she would probably back-off.

References:

[1] See Jerry Bergman, *Censoring the Darwin Skeptics: How Belief in Evolution is Enforced by Eliminating Dissidents,* Leafcutter Press, Southworth, WA, 2016;

Jerry Bergman, *Slaughter of the Dissidents: The Shocking Truth about Killing the Careers of Darwin Doubters,* Leafcutter Press, Southworth, WA, 2012

Chapter 6: Why Does Our Creator Care For Us?

Because our Creator's heart, mind, and capabilities are so much greater than ours, it is virtually impossible to thoroughly understand Him. However, He says He made us in His image and amazingly, we can communicate with Him at any time.

The wisdom He provides for us to live by, is for our most enjoyable life.

"I (Jesus) *have come that they may have life, and that they may have it more abundantly."* (John 10:10 NKJV)

We can observe His magnificent works all around us in His construction and maintenance of the beautiful people, foods, pets, flowers, birds, tropical fish, trees, mountains, waters, sun, moon, and stars.

Less obvious but equally as profound are His essential controls of gravity, electromagnetism, the strong and weak nuclear forces, electricity, light, and temperatures.

This God-based life-science of atomic biology reveals many of the details of His care for us every second of every day.

For all these awesome works done for us, there is no charge. Like the loving parent that He is, He keeps giving and giving to us, whether we thank Him or not. But, like any parent, He does expect some appreciation and recognition before we leave this world. He gives us a lifetime to acknowledge His enormous work and care for us. It is best to have Him as a close and appreciated generous friend, at all times rather than anger or ignore Him.

It is logical and understandable that He can become angry when someone or something else is given the credit for all of the constant and faithful work He performs for each of us every second of every day. The best-selling *Holy Bible* and other history books describe the troubles that come upon peoples and nations that disrespect Him and give credit for His enormous works to other causes from idols to evolutionism.

Among the many benefits and values of knowing our awesome Creator, Provider, and Maintainer is in cherishing the 'proof' of His obvious works, love, and care for each one of us.

This knowledge and understanding will be particularly beneficial for those students who learn it. The wisdom He provides in His best-selling 'operator's manual,' the Bible, has been of great benefit and value for centuries to those individuals and governments that have used it.

No scientist should be forced to pretend that there is no intelligence involved in the design and construction of living entities, including us.

One wonders how so many scientists who should know better, have been so obviously deceived or intimidated. There are, of course, many who just do not want to acknowledge or appreciate their essential, super-intelligent Creator and Provider, but they should not be allowed to keep the evidence away from others, especially students.

"The wrath of God is being revealed from heaven against all the godlessness and wickedness of men who suppress the truth ..." (Romans 1:18 NKJV)

Chapter 7: What Should Educators Do With This New Science of Atomic Biology?

As with all new concepts in science, atomic biology should be investigated impartially by individuals or committees that each group of educators deems reliable.

In this case, the term "impartial" is key because the topic of how living entities are constructed raises some emotional reactions, and emotions do not determine accuracy in science.

True science follows the evidence wherever it leads.

Through our three decades of research into what physical works with atoms have to be performed at each cell-construction site, we uncovered some very significant evidence. Our investigation of the 33 years of works by the three 2016 Nobel Prize Winners in Chemistry, provided the clinching evidence of the essentiality for a Super-Intelligent Cause in designing and building the 40 different kinds of molecular machines used in our 200+ different kinds of cells.

This was not the intent of the Nobel Winners, but it was a significant finding that derived from their work. Although they managed to build a few tiny molecular machines, the best units they could construct after 33 years of research and development are almost infinitely more simplistic than the simplest machines built for our new body cells every day of the week.

What they unintentionally 'proved' is that it takes far more intelligence than mankind has, to build the amazing molecular machines required for our cells.

As, by definition, evolution has no intelligence, evolution is now shown to be falsified as the true origin and cause of life.

As mentioned, when we asked evolutionists how they think the complex molecular machines for cells are made, those we have talked to declare that "natural mechanisms" do it. However, these natural mechanisms would have to be built containing far more intelligence and better equipment than the Nobel Prize Winners possessed and to work within the confines of a cell.

We have identified seven other principles and eighteen other essential works necessary to cause life that evolution is incapable of performing. These are listed in our full study/textbook titled,

Darwin's Replacement - How We Are Amazingly Made - Second Edition.

Our suggestion for education leaders is to, first, investigate atomic biology for accuracy, logic, and probability as the best explanation for the origin and cause of life.

If satisfied, then use this booklet to introduce the content to students in a simple three to five lesson life science course. It shows the evidence of how their awesome cell-parts and cells are carefully constructed by their Super-Intelligent and caring Creator, the God of their nation, in whom they can trust.

These booklets plus the full study/textbooks are available by contacting us at www.atomicbiology.com

Three to five lessons could provide students with significant and beneficial wisdom for a lifetime.

Should this God-based wisdom and knowledge be ignored and withheld, or shared with our students?

INDEX:

Contact Us:
For information on our textbooks, study books, courses, speakers, discounts, donations, participation, or other questions, please see our website: www.atomicbiology.com

Your support is invited to help bring our Creator, the God of our nations, back to our students. Please donate on our website.

Questions:

Chapter 1: How We Are Amazingly Made

1. What are two of the basic facts that have compelled scientists to move away from belief that "evolution" is the cause of life?

____

____

2. How many different types of cells have to be made in creating us? ______________________

3. About how many correctly selected, counted, and assembled atoms does it take to build each one of our red blood cells?

____

4. Because it takes about 6400 quadrillion correct atoms every second to make new red blood cells for a 150-pound person, about how many correct atoms are needed to make your new red blood cells? (xxx/150 lbs. x 6400 quadrillion) ________________ correct atoms every second 24/7.

5. To make the four bases for our DNA, what types of atoms are needed? ______________________________________

____

6. Do these atoms need to be carefully selected, counted, and assembled? ____________________

7. Where do these atoms come from? ___________________

____

8. In the diagram of a human eye, how many different eye parts are shown? ______________________

9. Besides seeing, what are the other four senses most people are blessed with? ____

____

10. About how many hours of recorded TV shows is our brain capable of storing? ___________________________________

11. Which explanation for the cause of life do you believe ("super-intelligent Creator" or "evolution") and why?

____

____

____

____

12. Do you think it is important to be taught what is true and accurate or does this not matter to you, and why?

____

____

Chapter 2: Why Is It "Super-Intelligence"?

1. How do we know that it takes a Super-Intelligent entity to build living things, including us?

____

____

____

2. Who or what is the only known entity with super-intelligence, far beyond mankind's level? ______________________________

3. Do our governments recognize this super-intelligent entity and how do they show it? ____________________________________

____

____

____

____

____

____

4. Where do the atoms come from to make our cell parts and cells? ____

____

5. According to Van Wedeen and L.L.Wald, about how many miles of fibers are made for our brains to connect the various regions?____

6. What does engineer-turned-molecular-biologist, Douglas Axe, believe is the most outstanding invention ever to exist and what is it made of? ____

____

7. How long does it take our Creator to convert a crawling worm into a flying butterfly? ______________________________

Chapter 3: Why Not "Evolution"?

1. According to the Theory of Evolution, an "universal common ancestor" came into being in some unknown way, with no intelligent help. What would this little entity have to be capable of doing all on its own? ______________________________

____

____

____

____

2. Because there is no common ancestor, what is there instead that can cause living things to exist? ____________________

____

3. Can atoms move themselves into their precise position in a cell? ______________________________________

4. Does the fossil record show the millions of transition entities needed for one kind of animal to evolve into another kind of animal?____

5. What if all the atoms and molecules for an entire animal with eyes, ears, nose, heart, brain, etc., were arranged precisely correctly, would anything else be needed for it to have life? ___

____

6. Is any intelligence needed to assemble our DNA? _________

7. Can 'evolution' which, by definition, has no intelligence to use, program the code necessary for our DNA? _____________

8. If evolution cannot do any of the super-intelligent works necessary to build a living entity, is it truly not the cause of life?

____

9. What are five of the other factors for life that require super-intelligence? ____

____

____

____

____

Chapter 4: God in Our Governments

1. What four countries are we looking at where God is highly recognized by their Government? __________________________

____

2. What three official national holidays that recognize God, are celebrated in these nations? ____________________________

____

3. What are five other ways these Governments officially recognize God? ____

____

____

4. Why does God deserve praise and recognition from our whole nation? ____

____

____

____

5. Do you believe that God has good advice for us to use in our life and can you give an example? ________________________

____

____

____

6. Regarding the "separation of church and state" concept, how do we know that God is not "the church"? __________________

____

____

____

____

7. Would you agree that God is a highly recognized part of some churches just as He is a highly recognized part of our Government?

____

____

____

Chapter 5: Why Is "Evolution" Still Being Taught?

1. What is the first reason 'evolution' is still being taught?

____

____

____

2. Why is this "anti-science"? ______________________________

____

____

____

____

3. What is the second reason 'evolution' is still being taught?

____

____

____

____

____

4. What are some of the problems caused by ignoring God and His advice? ______________________________

____

____

____

____

5. What new proof is there for the cause of life that should bring God back to our students? ______________________________

____

____

____

Chapter 6: Why Does Our Creator Care For Us?

1. In whose image are we made? ______________________________
2. Can we communicate with our Creator? ____________________

3. How can we observe His magnificent works? ___________

____

____

____

4. How much of the time does God work and care for us? ____

____

5. How much does He charge for all His work? _____________

6. What do you think He would like from us? _____________

____

____

____

7. Do you think God has the right to be upset and even angry with people who ignore Him and show no appreciation for all the work He does for us? ______________________________________

____

8. Where is His good advice written for us? _______________

9. Do you think it is right or wrong for science teachers to be forced to teach things that they know are not correct? _________

Chapter 7: What Should Educators Do With This New Science of Atomic Biology?

1. What do you think they should do? ____________________

____

____

____

CPSIA information can be obtained
at www.ICGtesting.com
Printed in the USA
LVHW071230120222
710985LV00027B/2074

9 781999 209773